GOD'S DESIGN® FOR HEAVEN & EARTH

OUR UNIVERSE
TEACHER SUPPLEMENT

1:1

answersingenesis

Petersburg, Kentucky, USA

ANSWERS IN GENESIS **SCIENCE** BY DEBBIE & RICHARD LAWRENCE

God's Design® for Heaven & Earth
Our Universe Teacher Supplement

© 2008 by Debbie & Richard Lawrence

Published by Answers in Genesis, 2800 Bullittsburg Church Rd., Petersburg KY 41080

You may contact the authors at (970) 686-5744.

ISBN: 1-60092-293-7

Cover design & layout: Diane King
Editors: Lori Jaworski, Gary Vaterlaus

Printed in China.

www.answersingenesis.org   www.godsdesignscience.com

TABLE OF CONTENTS

WELCOME TO
GOD'S DESIGN® FOR HEAVEN & EARTH

God's Design for Heaven and Earth is a series that has been designed for use in teaching earth science to elementary and middle school students. It is divided into three books: *Our Universe, Our Planet Earth,* and *Our Weather and Water.* Each book has 35 lessons including a final project that ties all of the lessons together.

In addition to the lessons, special features in each book include biographical information on interesting people as well as fun facts to make the subject more fun.

Although this is a complete curriculum, the information included here is just a beginning, so please feel free to add to each lesson as you see fit. A resource guide is included in the appendices to help you find additional information and resources. A list of supplies needed is included at the beginning of each lesson, while a master list of all supplies needed for the entire series can be found in the appendices.

Answer keys for all review questions, worksheets, quizzes, and the final exam are included here. Reproducible student worksheets and tests may be found on the supplementary CD-Rom for easy printing. Please contact Answers in Genesis if you wish to purchase a printed version of all the student materials, or go to www.AnswersBookstore.com.

If you wish to get through all three books of the *Heaven and Earth* series in one year, plan on covering approximately three lessons per week. The time required for each lesson varies depending on how much additional information you include, but plan on 20 minutes per lesson for beginners (grades 1–2) and 40 to 45 minutes for grades 3–8.

Quizzes may be given at the conclusion of each unit and the final exam may be given after lesson 34.

If you wish to cover the material in more depth, you may add additional information and take a longer period of time to cover all the material, or you could choose to do only one or two of the books in the series as a unit study.

WHY TEACH EARTH SCIENCE?

It is not uncommon to question the need to teach children hands-on science in elementary or middle school. We could argue that the knowledge gained in science will be needed later in life in order for children to be more productive and well-rounded adults. We could argue that teaching children science also teaches them logical and inductive thinking and reasoning skills, which are tools they will need to be more successful. We could argue that science is a necessity in this technological world

in which we live. While all of these arguments are true, not one of them is the main reason that we should teach our children science. The most important reason to teach science in elementary school is to give children an understanding that God is our Creator, and the Bible can be trusted. Teaching science from a creation perspective is one of the best ways to reinforce our children's faith in God and to help them counter the evolutionary propaganda they face every day.

God is the Master Creator of everything. His handiwork is all around us. Our great Creator put in place all of the laws of physics, biology, and chemistry. These laws were put here for us to see His wisdom and power. In science, we see the hand of God at work more than in any other subject. Romans 1:20 says, "For since the creation of the world His invisible attributes are clearly seen, being understood by the things that are made, even His eternal power and Godhead, so that they [men] are without excuse." We need to help our children see God as Creator of the world around them so they will be able to recognize God and follow Him.

The study of earth science helps us to understand and appreciate this amazing world God gave us. Studying the processes that shape the earth, and exploring the origins of the earth and the universe often bring us into direct conflict with evolutionary theories. This is why it is so critical to teach our children the truth of the Bible, how to evaluate the evidence, how to distinguish fact from theory, and to realize that the evidence, rightly interpreted, supports biblical creation not evolution.

It's fun to teach earth science! It's interesting too. Rocks, weather, and stars are all around us. Children naturally collect rocks and gaze at the stars. You just need to direct their curiosity.

Finally, teaching earth science is easy. It's where you live. You won't have to try to find strange materials for experiments or do dangerous things to learn about the earth.

HOW DO I TEACH SCIENCE?

In order to teach any subject you need to understand how people learn. People learn in different ways. Most people, and children in particular, have a dominant or preferred learning style in which they absorb and retain information more easily.

If a student's dominant style is:

AUDITORY
He needs not only to hear the information but he needs to hear himself say it. This child needs oral presentation as well as oral drill and repetition.

VISUAL
She needs things she can see. This child responds well to flashcards, pictures, charts, models, etc.

KINESTHETIC
he needs active participation. This child remembers best through games, hands-on activities, experiments, and field trips.

Also, some people are more relational while others are more analytical. The relational student needs to know why this subject is important, and how it will affect him personally. The analytical student, however, wants just the facts.

If you are trying to teach more than one student, you will probably have to deal with more than one learning style. Therefore, you need to present your lessons in several different ways so that each student can grasp and retain the information.

GRADES 1–2

Because *God's Design Science* books are designed to be used with students in grades 1–8, each lesson has been divided into three sections. The "Beginner" section is for students in grades 1–2. This part contains a read-aloud section explaining the material for that lesson followed by a few questions to make sure that the students understand what they just heard. We recommend that you do the hands-on activity in the blue box in the main part of the lesson to help your students see and understand the concepts.

GRADES 3–8

The second part of each lesson should be completed by all upper elementary and junior high students. This is the main part of the lesson containing a reading section, a hands-on activity that reinforces the ideas in the reading section (blue box), and a review section that provides review questions and application questions (red box).

GRADES 6–8

Finally, for middle school/junior high age students, we provide a "Challenge" section that contains more challenging material as well as additional activities and projects for older students (green box).

We have included periodic biographies to help your students appreciate the great men and women who have gone before us in the field of science.

We suggest a threefold approach to each lesson:

INTRODUCE THE TOPIC

We give a brief description of the facts. Frequently you will want to add more information than the essentials given in this book. In addition to reading this section aloud (or having older children read it on their own), you may wish to do one or more of the following:

- Read a related book with your students.
- Write things down to help your visual learners.
- Give some history of the subject. We provide some historical sketches to help you, but you may want to add more.
- Ask questions to get your students thinking about the subject.

MAKE OBSERVATIONS AND DO EXPERIMENTS

- Hands-on projects are suggested for each lesson. This part of each lesson may require help from the teacher.
- Have your students perform the activity by themselves whenever possible.

REVIEW

- The "What did we learn?" section has review questions.
- The "Taking it further" section encourages students to
  - Draw conclusions
  - Make applications of what was learned
  - Add extended information to what was covered in the lesson
- The "FUN FACT" section adds fun or interesting information.

By teaching all three parts of the lesson, you will be presenting the material in a way that children with any learning style can both relate to and remember.

Also, this approach relates directly to the scientific method and will help your students think more scientifically. The *scientific method* is just a way to examine a subject logically and learn from it. Briefly, the steps of the scientific method are:

1. Learn about a topic.
2. Ask a question.
3. Make a hypothesis (a good guess).
4. Design an experiment to test your hypothesis.
5. Observe the experiment and collect data.
6. Draw conclusions. (Does the data support your hypothesis?)

Note: It's okay to have a "wrong hypothesis." That's how we learn. Be sure to help your students understand why they sometimes get a different result than expected.

Our lessons will help your students begin to approach problems in a logical, scientific way.

HOW DO I TEACH CREATION VS. EVOLUTION?

We are constantly bombarded by evolutionary ideas about the earth in books, movies, museums, and even commercials. These raise many questions: What is the big bang? How old is the earth? Do fossils show evolution to be true? Was there really a worldwide flood? When did dinosaurs live? Was there an ice age? How can we teach our children the truth about the origins of the earth? The Bible answers these questions and this book accepts the historical accuracy of the Bible as written. We believe this is the only way we can teach our children to trust that everything God says is true.

There are five common views of the origins of life and the age of the earth:

Historical biblical account	Progressive creation	Gap theory	Theistic evolution	Naturalistic evolution
Each day of creation in Genesis is a normal day of about 24 hours in length, in which God created everything that exists. The earth is only thousands of years old, as determined by the genealogies in the Bible.	The idea that God created various creatures to replace other creatures that died out over millions of years. Each of the days in Genesis represents a long period of time (day-age view) and the earth is billions of years old.	The idea that there was a long, long time between what happened in Genesis 1:1 and what happened in Genesis 1:2. During this time, the "fossil record" was supposed to have formed, and millions of years of earth history supposedly passed.	The idea that God used the process of evolution over millions of years (involving struggle and death) to bring about what we see today.	The view that there is no God and evolution of all life forms happened by purely naturalistic processes over billions of years.

Any theory that tries to combine the evolutionary time frame with creation presupposes that death entered the world before Adam sinned, which contradicts what God has said in His Word. The view that the earth (and its "fossil record") is hundreds of millions of years old damages the gospel message. God's completed creation was "very good" at the end of the sixth day (Genesis 1:31). Death entered this perfect paradise *after* Adam disobeyed God's command. It was the punishment for Adam's sin (Genesis 2:16–17, 3:19; Romans 5:12–19). Thorns appeared when God cursed the ground because of Adam's sin (Genesis 3:18).

The first animal death occurred when God killed at least one animal, shedding its blood, to make clothes for Adam and Eve (Genesis 3:21). If the earth's "fossil record" (filled with death, disease, and thorns) formed over millions of years before Adam appeared (and before he sinned), then death no longer would be the penalty for sin. Death, the "last enemy" (1 Corinthians 15:26), diseases (such as cancer), and thorns would instead be part of the original creation that God labeled "very good." No, it is clear that the "fossil record" formed some time *after* Adam sinned—not many millions of years before. Most fossils were formed as a result of the worldwide Genesis Flood.

When viewed from a biblical perspective, the scientific evidence clearly supports a recent creation by God, and not naturalistic evolution and millions of years. The volume of evidence supporting the biblical creation account is substantial and cannot be adequately covered in this book. If you would like more information on this topic, please see the resource guide in Appendix A. To help get you started, just a few examples of evidence supporting biblical creation are given below:

Evolutionary Myth: The earth is 4.6 billion years old.

The Truth: Many processes observed today point to a young earth of only a few thousand years. The rate at which the earth's magnetic field is decaying suggests the earth must be less than 10,000 years old. The rate of population growth and the recent emergence of civilization suggests only a few thousand years of human population. And, at the current rate of accumulation, the amount of mud on the sea floor should be many kilometers thick if the earth were billions of years old. However, the average depth of all the mud in the whole ocean is less than 400 meters, giving a maximum age for the earth of not more than 12 million years. All this and more indicates an earth much younger than 4.6 billion years.

John D. Morris, *The Young Earth* (Creation Life Publishers, 1994), pp. 70–71, 83–90. See also "Get Answers: Young Age Evidence" at www.answersingenesis.org/go/young.

Evolutionary Myth: The universe formed from the big bang.

The Truth: There are many problems with this theory. It does not explain where the initial material came from. It cannot explain what caused that material to fly apart in the first place. And nothing in physics indicates what would make the particles begin to stick together instead of flying off into space forever. The big bang theory contradicts many scientific laws. Because of these problems, some scientists have abandoned the big bang and are attempting to develop new theories to explain the origin of the universe.

Jason Lisle, "Does the Big Bang Fit with the Bible," in *The New Answers Book 2*, Ken Ham, ed. (Master Books, 2008). See also "What are some of the problems with the big bang hypothesis?" at www.answersingenesis.org/go/big-bang.

Evolutionary Myth: Fossils prove evolution.

The Truth: While Darwin predicted that the fossil record would show numerous transitional fossils, even more than 145 years later, all we have are a handful of disputable examples. For example, there are no fossils showing something that is part way between a dinosaur and a bird. Fossils show that a snail has always been a snail; a squid has always been a squid. God created each animal to reproduce after its kind (Genesis 1:20–25).

Evolutionary Myth: There is not enough water for a worldwide flood.

The Truth: Prior to the Flood, just as today, much of the water was stored beneath the surface of the earth. In addition, Genesis 1 states that the water below was separated from the water above, indicating that the atmosphere may have contained a great deal more water than it does today. Also, it is likely that before the Flood the mountains were not as high as they are today, but that the mountains rose and the valleys sank *after* the Flood began, as Psalm 104:6–9 suggests. At the beginning of the Flood, the fountains of the deep burst forth and it rained for 40 days and nights. This could have provided more than enough water to flood the entire earth. Indeed, if the entire earth's surface were leveled by smoothing out the topography of not only the land surface but also the rock surface on the ocean floor, the waters of the present-day oceans would cover the earth's surface to a depth of 1.7 miles (2.7 kilometers). Fossils have been found on the highest mountain peaks around the world showing that the waters of the Flood did indeed cover the entire earth.

Ken Ham & Tim Lovett, "What There Really a Noah's Ark and Flood," in *The New Answers Book 1*, Ken Ham, ed. (Master Books, 2006).

Evolutionary Myth: Slow climate changes over time have resulted in multiple ice ages.

The Truth: There is widespread evidence of glaciers in many parts of the world indicating one ice age. Evolutionists find the cause of the Ice Age a mystery. Obviously, the climate would need to be colder. But global cooling by itself is not enough, because then there would be less evaporation, so less snow. How is it possible to have both a cold climate and lots of evaporation? The Ice Age was most likely an aftermath of Noah's Flood. When "all the fountains of the great deep" broke up, much hot water and lava would have poured directly into the oceans. This would have warmed the oceans, increasing evaporation. At the same time, much volcanic ash in the air after the Flood would have blocked out much sunlight, cooling the land. So the Flood would have produced the necessary combination of increased evaporation from the warmed oceans and cool continental climate from the volcanic ash in the air. This would have resulted in increased snowfall over the continents. With the snow falling faster than it melted, ice sheets would have built up. The Ice Age probably lasted less than 700 years.

Michael Oard, *Frozen in Time* (Master Books, 2004). See also www.answersingenesis.org/go/ice-age.

Evolutionary Myth: Thousands of random changes over millions of years resulted in the earth we see today.

The Truth: The second law of thermodynamics describes how any system tends toward a state of zero entropy or disorder. We observe how everything around us becomes less organized and loses energy. The changes required for the formation of the universe, the planet earth and life, all from disorder, run counter to the physical laws we see at work today. There is no known mechanism to harness the raw energy of the universe and generate the specified complexity we see all around us.

John D. Morris, *The Young Earth* (Creation Life Publishers, 1994), p. 43. See also www.answersingenesis.org/go/thermodynamics.

Despite the claims of many scientists, if you examine the evidence objectively, it is obvious that evolution and millions of years have not been proven. You can be confident that if you teach that what the Bible says is true, you won't go wrong. Instill in your student a confidence in the truth of the Bible in all areas. If scientific thought seems to contradict the Bible, realize that scientists often make mistakes, but God does not lie. At one time scientists believed that the earth was the center of the universe, that living things could spring from non-living things, and that blood-letting was good for the body. All of these were believed to be scientific facts but have since been disproved, but the Word of God remains true. If we use modern "science" to interpret the Bible, what will happen to our faith in God's Word when scientists change their theories yet again?

INTEGRATING THE SEVEN C'S

Throughout the *God's Design® for Science* series you will see icons that represent the Seven C's of History. The Seven C's is a framework in which all of history, and the future to come, can be placed. As we go through our daily routines we may not understand how the details of life connect with the truth that we find in the Bible. This is also the case for students. When discussing the importance of the Bible you may find yourself telling students that the Bible is relevant in everyday activities. But how do we help the younger generation see that? The Seven C's are intended to help.

The Seven C's can be used to develop a biblical worldview in students, young or old. Much more than entertaining stories and religious teachings, the Bible has real connections to our everyday life. It may be hard, at first, to see how many connections there are, but with practice ,the daily relevance of God's Word will come alive. Let's look at the Seven C's of History and how each can be connected to what the students are learning.

work of God. Virtually all of the lessons in *God's Design for Science* can be related to God's creation of the heavens and earth.

Other contexts include:

Natural laws—any discussion of a law of nature naturally leads to God's creative power.

DNA and information—the information in every living thing was created by God's supreme intelligence.

Mathematics—the laws of mathematics reflect the order of the Creator.

Biological diversity—the distinct kinds of animals that we see were created during the Creation Week, not as products of evolution.

Art—the creativity of man is demonstrated through various art forms.

History—all time scales can be compared to the biblical time scale extending back about 6,000 years.

Ecology—God has called mankind to act as stewards over His creation.

CREATION

God perfectly created the heavens, the earth, and all that is in them in six normal-length days around 6,000 years ago.

This teaching is foundational to a biblical worldview and can be put into the context of any subject. In science, the amazing design that we see in nature—whether in the veins of a leaf or the complexity of your hand—is all the handi-

CORRUPTION

After God completed His perfect creation, Adam disobeyed God by eating the forbidden fruit. As a result, sin and death entered the world, and the world has been in decay since that time. This point is evident throughout the world that we live in. The struggle for survival in animals, the death of loved ones, and the violence all around us are all examples of the corrupting influence of sin.

Other contexts include:

Genetics—the mutations that lead to diseases, cancer, and variation within populations are the result of corruption.

Biological relationships—predators and parasites result from corruption.

History—wars and struggles between mankind, exemplified in the account of Cain and Abel, are a result of sin.

CATASTROPHE

God was grieved by the wickedness of mankind and judged this wickedness with a global Flood. The Flood covered the entire surface of the earth and killed all air-breathing creatures that were not aboard the Ark. The eight people and the animals aboard the Ark replenished the earth after God delivered them from the catastrophe.

The catastrophe described in the Bible would naturally leave behind much evidence. The studies of geology and of the biological diversity of animals on the planet are two of the most obvious applications of this event. Much of scientific understanding is based on how a scientist views the events of the Genesis Flood.

Other contexts include:

Biological diversity—all of the birds, mammals, and other air-breathing animals have populated the earth from the original kinds which left the Ark.

Geology—the layers of sedimentary rock seen in roadcuts, canyons, and other geologic features are testaments to the global Flood.

Geography—features like mountains, valleys, and plains were formed as the floodwaters receded.

Physics—rainbows are a perennial sign of God's faithfulness and His pledge to never flood the entire earth again.

Fossils—Most fossils are a result of the Flood rapidly burying plants and animals.

Plate tectonics—the rapid movement of the earth's plates likely accompanied the Flood.

Global warming/Ice Age—both of these items are likely a result of the activity of the Flood. The warming we are experiencing today has been present since the peak of the Ice Age (with variations over time).

CONFUSION

God commanded Noah and his descendants to spread across the earth. The refusal to obey this command and the building of the tower at Babel caused God to judge this sin. The common language of the people was confused and they spread across the globe as groups with a common language. All people are truly of "one blood" as descendants of Noah and, originally, Adam.

The confusion of the languages led people to scatter across the globe. As people settled in new areas, the traits they carried with them became concentrated in those populations. Traits like dark skin were beneficial in the tropics while other traits benefited populations in northern climates, and distinct people groups, not races, developed.

Other contexts include:

Genetics—the study of human DNA has shown that there is little difference in the genetic makeup of the so-called "races."

Languages—there are about seventy language groups from which all modern languages have developed.

Archaeology—the presence of common building structures, like pyramids, around the world confirms the biblical account.

Literature—recorded and oral records tell of similar events relating to the Flood and the dispersion at Babel.

CHRIST

God did not leave mankind without a way to be redeemed from its sinful state. The Law was given to Moses to show how far away man is from God's standard of perfection. Rather than the sacrifices, which only covered sins, people needed a Savior to take away their sin. This was accomplished when Jesus Christ came to earth to live a perfect life and, by that obedience, was able to be the sacrifice to satisfy God's wrath for all who believe.

The deity of Christ and the amazing plan that was set forth before the foundation of the earth is the core of Christian doctrine. The earthly life of Jesus was the fulfillment of many prophecies and confirms the truthfulness of the Bible. His miracles and presence in human form demonstrate that God is both intimately concerned with His creation and able to control it in an absolute way.

Other contexts include:

Psychology—popular secular psychology teaches of the inherent goodness of man, but Christ has lived the only perfect life. Mankind needs a Savior to redeem it from its unrighteousness.

Biology—Christ's virgin birth demonstrates God's sovereignty over nature.

Physics—turning the water into wine and the feeding of the five thousand demonstrate Christ's deity and His sovereignty over nature.

History—time is marked (in the western world) based on the birth of Christ despite current efforts to change the meaning.

Art—much art is based on the life of Christ and many of the masters are known for these depictions, whether on canvas or in music.

CROSS

Because God is perfectly just and holy, He must punish sin. The sinless life of Jesus Christ was offered as a substitutionary sacrifice for all of those who will repent and put their faith in the Savior. After His death on the Cross, He defeated death by rising on the third day and is now seated at the right hand of God.

The events surrounding the crucifixion and resurrection have a most significant place in the life of Christians. Though there is no way to scientifically prove the resurrection, there is likewise no way to prove the stories of evolutionary history. These are matters of faith founded in the truth of God's Word and His character. The eyewitness testimony of over 500 people and the written Word of God provide the basis for our belief.

Other contexts include:

Biology—the biological details of the crucifixion can be studied alongside the anatomy of the human body.

History—the use of crucifixion as a method of punishment was short-lived in historical terms and not known at the time it was prophesied.

Art—the crucifixion and resurrection have inspired many wonderful works of art.

CONSUMMATION

God, in His great mercy, has promised that He will restore the earth to its original state—a world without death, suffering, war, and disease. The corruption introduced by Adam's sin will be removed. Those who have repented and put their trust in the completed work of Christ on the Cross will experience life in this new heaven and earth. We will be able to enjoy and worship God forever in a perfect place.

This future event is a little more difficult to connect with academic subjects. However, the hope of a life in God's presence and in the absence of sin can be inserted in discussions of human conflict, disease, suffering, and sin in general.

Other contexts include:

History—in discussions of war or human conflict the coming age offers hope.

Biology—the violent struggle for life seen in the predator-prey relationships will no longer taint the earth.

Medicine—while we struggle to find cures for diseases and alleviate the suffering of those enduring the effects of the Curse, we ultimately place our hope in the healing that will come in the eternal state.

The preceding examples are given to provide ideas for integrating the Seven C's of History into a broad range of curriculum activities. We would recommend that you give your students, and yourself, a better understanding of the Seven C's framework by using AiG's *Answers for Kids* curriculum.

The first seven lessons of this curriculum cover the Seven C's and will establish a solid understanding of the true history, and future, of the universe. Full lesson plans, activities, and student resources are provided in the curriculum set.

We also offer bookmarks displaying the Seven C's and a wall chart. These can be used as visual cues for the students to help them recall the information and integrate new learning into its proper place in a biblical worldview.

Even if you use other curricula, you can still incorporate the Seven C's teaching into those. Using this approach will help students make firm connections between biblical events and every aspect of the world around them, and they will begin to develop a truly biblical worldview and not just add pieces of the Bible to what they learn in "the real world."

SPACE MODELS & TOOLS

LESSON 1

INTRODUCTION TO ASTRONOMY

STUDY OF SPACE

SUPPLY LIST

Bible   Copy of "God's Purpose for the Universe" worksheet

Supplies for Challenge: Copy of "Knowledge of the Stars" worksheet

BEGINNERS

- What is astronomy? **The study of the things in space.**

GOD'S PURPOSE FOR THE UNIVERSE WORKSHEET

1. I was designed to rule the day: **Sun/greater light.**

2. I was designed to rule the night: **Moon/lesser light.**

3. We are times that are to be marked by the movement of the sun, moon, and stars: **Seasons, days, and years.**

4. Besides marking times, I am another reason why the sun, moon, and stars were made: **To give light and to show signs.**

5. We were made by God's hands and this is what will eventually happen to us: **Heavens and earth will perish and wear out.**

6. This is higher than me (the earth): **The heavens.**

7. I am what you will see in the heavens in the last days: **Wonders, sun to darkness, moon to blood.**

8. I (the sun), stood still for this long, until Joshua and the Israelites defeated their enemies: **About a full day.**

WHAT DID WE LEARN?

- What is astronomy? **The study of the stars, planets, moons, and other items in space.**

- Why should we want to study astronomy? **To learn more about God's creation and see His glory.**

TAKING IT FURTHER

- What is one thing you really want to learn during this study? **Answers will vary.**

- Write your question or questions on a piece of paper and save it to make sure you find the answers by the end of the book. **Encourage the student to do this and keep it in an accessible place.**

CHALLENGE: KNOWLEDGE OF THE STARS WORKSHEET

1. What is the nearest star to the earth? **Sun.**

2. What are the main elements in stars? **Hydrogen and helium.**

3. What is the name of the galaxy that we live in? **Milky Way.**

4. What is special about Polaris, the North Star? **It does not appear to move through the sky like the other stars.**

5. What unit of distance is used to measure items in space? **Light year, parsec, or astronomical unit.**

6. What name describes when one celestial body blocks the light from another? **Eclipse.**

7. What force holds the planets in their places? **Gravity.**

8. Name three items found in space besides stars, moons, and planets. **Comets, asteroids, meteors, plutoids, space junk, satellites, space station.**

9. Name two scientists important to our understanding of astronomy. **Newton, Galileo, Copernicus, Kepler, Hubble.**

10. How long does it take for light to travel from the sun to the earth? **About 8 minutes.**

LESSON 2

Space Models

What's really out there?

Supply list

Book   Piece of paper   Ping-pong ball   Golf ball

Beginners

- Does the sun move around the earth or does the earth move around the sun? **The earth moves around the sun.**

- What force keeps all the planets, moons, and stars in their places? **Gravity.**

What did we learn?

- What are the two major models that have been used to describe the arrangement of the universe? **Geocentric/Ptolemaic—earth centered, and Heliocentric/Copernican—sun centered.**

- What was the main idea of the Geocentric Model? **The earth was the center of the universe and everything revolved around it.**

- What is the main idea of the Heliocentric Model? **The sun is the center of the solar system and the earth and other planets revolve around it.**

- What force holds all of the planets in orbit around the sun? **Gravity.**

Taking it further

- Which exerts the most gravitational pull, the earth or the sun? **The sun because it is much more massive than the earth.**

- If the sun has a stronger gravitational pull, then why aren't objects pulled off of the earth toward the sun? **The strength of the gravitational pull decreases with distance. The pull of the earth is stronger on us because we are so much closer to the center of the earth than we are to the sun. If an object moves far enough away from the earth, the earth's gravity no longer has much effect on it. And if that object moves close enough to the sun, it will be pulled into the sun by the sun's gravity.**

THE EARTH'S MOVEMENT

ROTATING AND REVOLVING

SUPPLY LIST

Flashlight   Basketball or volleyball   Masking tape

Supplies for Challenge: Copy of "Clock" pattern   Tripod   Thread   Needle

Modeling clay   Turntable, swivel chair or stool, Lazy Susan, etc.

BEGINNERS

- In what two ways is the earth moving? **Rotating on its axis and orbiting or revolving around the sun.**

- Why do we experience seasons like summer and winter? **The earth is tilted with respect to the sun. When your part of the earth is tilted toward the sun it is summer, and when it is tilted away from the sun it is winter.**

WHAT DID WE LEARN?

- What are the two different types of motion that the earth experiences? **Rotation on its axis and revolution around the sun.**

- What observations can we make that are the result of the rotation of the earth on its axis? **Day and night, the stars rotating in the sky, the bulging of the earth, diagonal air flow.**

- What observations can we make that are the result of the revolution of the earth around the sun? **Changing of the seasons, parallax of stars, more meteors observed after midnight.**

- What is a solstice? **The first day of summer or the first day of winter, when the earth is in the place in its orbit where the sun is hitting directly on either the Tropic of Cancer or the Tropic of Capricorn.**

- What is an equinox? **The first day of spring or the first day of autumn, when the earth is halfway between the solstices.**

TAKING IT FURTHER

- What are the advantages of the earth being tilted on its axis as it revolves around the sun? **This gives us seasons. Without this tilt, the temperatures would be relatively stable year round. This would result in less of the earth being able to grow crops. Only the warm areas near the equator would have warm enough weather to grow food.**

- One argument against Copernicus's theory was that if the earth were moving, flying birds would be left behind. Why don't the birds get left behind as the earth moves through space? **The atmosphere in which the birds are flying moves with the earth because of gravity.**

CHALLENGE: FOUCAULT PENDULUM

- What forces are affecting the pendulum? **Gravity is pulling down on the weight at the end of the pendulum, and air is resisting the movement of the pendulum.**

- Why does the pendulum eventually stop moving? **Because of the air resistance.**

- How does a Foucault pendulum keep moving for hours or days at a time without stopping? **They are often designed with an iron ring near the top where the pendulum is attached to the building. Also, there are electromagnets placed around the ring. As the pendulum swings through a certain part of its arc the magnet turns on, attracting the ring. Then the magnet turns off to allow the pendulum to swing freely. This magnet system compensates for the air resistance that the pendulum experiences, so it does not slow down.**

LESSON 4

TOOLS FOR STUDYING SPACE

DO I NEED MORE THAN MY EYES?

SUPPLY LIST

Mirror   Magnifying glass   Flashlight   Optional: Access to a telescope
Supplies for Challenge: Paper   Marker   Car

BEGINNERS

- What is a telescope and what is it used for? **An instrument with lenses and mirrors that makes something look bigger. It is used to view distant objects like stars, planets, and other things in space.**

WHAT DID WE LEARN?

- What are the three main types of telescopes? **Refracting—using only lenses, Reflecting—using lenses and mirrors, Radio—collecting radio waves.**
- What was one disadvantage of the early refracting telescope? **The refracting lens bent the light causing false colors to appear around the edges of the image.**
- How did Newton avoid this problem? **He used mirrors instead of a lens to collect the light.**

TAKING IT FURTHER

- Why do you think scientists wanted to put a telescope in space? **It would give better images without the interference of the earth's atmosphere. It would also not be affected by the movement of the earth and it would not be limited to one position on the earth.**
- What kinds of things can we learn from using optical telescopes? **What a star looks like, its color, brilliance, etc. Astronomers have observed that occasionally what appears to be only one star might actually be two stars.**
- What kinds of things can we learn from radio telescopes? **The radio wave activity of a star or other object in space can be detected. Different substances emit different wavelengths of radio waves so the composition of a star can be determined. Also, using radio waves as radar allows us to get an idea of the density of planets.**

QUIZ 1

SPACE MODELS & TOOLS

LESSONS 1–4

Short answer:

1. What are the two ways the earth moves in space? **Rotates on axis, revolves around sun.**
2. Why does the earth experience seasons? **Its axis is tilted with respect to the sun.**
3. Why are seasons an indication of God's provision for man? **The seasons allow more of the earth to be cultivated for food.**
4. What is the main idea of the Geocentric Model? **The earth is the center of our solar system/universe.**
5. What is the main idea of the Heliocentric Model? **The sun is the center of our solar system.**

Mark each statement either True or False.

6. _F_ Scientists can prove where the earth came from.

7. _F_ Scientists have proven the Big Bang is true.

8. _T_ The Bible can tell us some things about astronomy.

9. _F_ The Bible can tell us everything about astronomy.

10. _T_ Gravity is the force that holds all planets in orbit.

11. _F_ Galileo invented the first telescope.

12. _T_ Heavier objects exert more gravity than lighter objects.

13. _T_ Closer objects exert more gravity than ones farther away.

14. _F_ The sun exerts more gravity on us than the earth does.

15. _T_ The Bible says that God's power can be seen in His creation.

16. _T_ Newton's reflecting telescope reduced chromatic aberration.

CHALLENGE QUESTIONS

Short answer:

17. List at least one contribution that each of the following men made to the study of astronomy:

    Nicolaus Copernicus: **Developed the Heliocentric Model of the solar system.**

    Galileo: **First to use a telescope to study the heavens, discovered moons around Jupiter and rings around Saturn, supported Heliocentric Model.**

    Sir Isaac Newton: **Defined the laws of gravitation, developed a reflecting telescope to reduce chromatic aberration.**

18. What invention did you study that demonstrates the rotation of the earth? **Foucault pendulum.**

19. Why do scientists look for ways to make telescopes larger? **Larger telescopes gather more light and give brighter, clearer images.**

20. Explain briefly how the mirror of the Keck telescope is made? **The Keck telescope mirror consists of many hexagon shaped mirrors that fit together to form a very large mirror.**

LESSON 5

OVERVIEW OF THE UNIVERSE

HOW BIG IS IT?

SUPPLY LIST

Star chart   Clear night sky

BEGINNERS

- What objects make up our solar system? **The sun and all the things that orbit it.**
- What is a constellation? **A collection of stars that make a particular picture.**
- Why did sailors need to be able to recognize stars? **So they could locate them in the night sky and use them for navigation.**

WHAT DID WE LEARN?

- What is our solar system? **The group of heavenly bodies that includes our sun and the planets that revolve around it.**
- Our solar system is part of which galaxy? **The Milky Way galaxy.**
- How big is the universe? **No one knows for sure. Some people believe that is has no end.**

TAKING IT FURTHER

- Why do you think our galaxy is called the Milky Way? **Because on clear nights the stars in the galaxy make a white milky band across the sky.**
- Why do you need star charts that are different for different times of the year? **Because as the earth travels around the sun, it is in a slightly different position with respect to the stars each day.**
- Why do you need star charts that are different for different times of the night? **Because as the earth rotates on its axis, a particular spot on the earth moves with respect to the stars.**

CHALLENGE: LOCATING STARS

- Explain how a star map is similar to a map of the globe. **They both have an equator and prime longitude line. They both allow you to locate areas using the equivalent of latitude and longitude.**
- What units are used to measure declination and ascension? **Declination—degrees north or south, ascension—hours and minutes.**
- How does an astronomer define a constellation differently than most people? **To an astronomer, a constellation is an area in the sky; to most people, a constellation is a collection of stars that forms a picture of sorts.**

LESSON 6

STARS

TWINKLE, TWINKLE LITTLE STAR

SUPPLY LIST

Copy of "Starlight" worksheet   Ruler and a yardstick
2 flashlights (one brighter than the other or with different sizes of lenses)
Supplies for Challenge: Calculator

BEGINNERS

- Why might some stars look different from other stars? **They are different colors; some are farther away; they are made from different materials; some are brighter.**

STARLIGHT WORKSHEET

- What happened to the light beam as the flashlight was moved farther from the wall? **The beam got wider but was not as intense; the closer one would appear brighter.**

- How would two identical stars appear to someone on earth if one was much farther away? **The one farther away would appear smaller and dimmer than the closer one.**

- Why could two stars with the same apparent brightness be different distances from the earth? **Because the way the stars appear to us is determined by the brightness of the star and the distance to the star; if one star was much brighter than the other but also farther away, the same amount of light could be hitting the earth from both stars. This would make them appear the same to us.**

WHAT DID WE LEARN?

- What is the unit of distance used to measure how far away something is in space? **Light-year.**

- How far is a light-year? **The distance light travels in one year—about 6 trillion miles.**

- What does the color of a star tell us about that star? **Its approximate surface temperature; blue stars are much hotter than yellow or red stars.**

TAKING IT FURTHER

- What causes stars to appear to move in the sky? **Most apparent motion is caused by the movement of the earth.**

- How can we determine if a star's absolute distance from the earth is actually changing over time? **We must measure its light over a long period of time and see if it is changing.**

- Why is brightness not a good indicator of the distance of a star from the earth? **Brightness is determined by the amount of light emitted and how far the star is from the earth. Brighter stars may be farther away but emitting more light, or they may be closer and emitting less light. You need to know how much light is being emitted, as well as the brightness, to determine the distance of the star from earth.**

LESSON 7

HEAVENLY BODIES

MORE THAN JUST STARS

SUPPLY LIST

Flashlight   Pencil

BEGINNERS

- What is a galaxy? **A large group of stars that rotates together.**
- What is the name of our galaxy? **The Milky Way.**

WHAT DID WE LEARN?

- What is a cluster of stars? **A group of stars that appear to move together.**
- What is a galaxy? **A group of millions (or billions) of stars that rotates around a central point.**
- Explain the difference between a nova, a supernova, and a neutron star. **A nova is a star that is exploding and then returns to normal. A supernova is a star that experiences such a huge explosion that it may be destroyed. A neutron star is believed to be what is left of a supernova. It is extremely small and dense and emits radio waves.**

TAKING IT FURTHER

- How can a star appear to become brighter and dimmer on a regular basis? **If two stars rotate around each other, they can line up so that they appear as one bright star. Later, one star can block the light of the other making it appear dimmer. Also, some stars expand and contract on a regular basis causing them to appear brighter and dimmer—these are called Cepheid variable stars.**
- Why does starlight from millions of light-years away not prove that the earth is old? **There are several ideas using general relativity that explain how time may have passed more slowly on earth while billions of years were passing on the stars in the expanding universe.**

LESSON 8

ASTEROIDS

MINOR PLANETS

SUPPLY LIST

Paper and pencil
Supplies for Challenge: Research materials for the Trojan War

BEGINNERS

- What is an asteroid? **A large piece of rock orbiting the sun.**
- Where is the asteroid belt located? **Between the orbits of Mars and Jupiter.**

WHAT DID WE LEARN?

- What is an asteroid? **A relatively small rock in a regular orbit around the sun.**
- Where are most asteroids in our solar system located? **In the asteroid belt between the orbits of Mars and Jupiter.**
- What is another name for asteroids? **Minor planets.**

TAKING IT FURTHER

- What is the chance that an asteroid will hit the earth? **Relatively small. As Christians we must trust God that all things, including asteroids, are in His control.**

CHALLENGE: TROJAN ASTEROIDS

- **Some of the names of the Trojan asteroids include: Achilles, Hektor, Nestor, Agamemnon, Odysseus, Ajax, Diomedes, Antilochus, and Menelaus.**

LESSON 9

COMETS

LOOK AT THAT TAIL!

SUPPLY LIST

Small Styrofoam ball   Tagboard/poster board   Glue   Glitter

BEGINNERS

- What is a comet? **A ball of ice and dust that orbits the sun.**
- What are the two parts of a comet? **The head and the tail.**

WHAT DID WE LEARN?

- What is a comet? **A frozen core of rock and dust that orbits the sun in a regular orbit.**
- Who was the first person to accurately predict the orbit of comets? **Edmond Halley.**
- What are the two main parts of a comet? **Head, tail.**

TAKING IT FURTHER

- Why does a comet's tail always point away from the sun? **The solar winds that cause the tail are always moving away from the sun.**
- Why doesn't a comet have a tail when it is far from the sun? **There are no solar winds to push it, and it does not vaporize when it is far from the sun.**
- When will Halley's Comet next appear? **1986+75 = 2061.**

CHALLENGE: GOD CREATED COMETS

1. **Comets become smaller every time they pass the sun. Thus they are wearing out just as the earth and the rest of the universe is wearing out.**
2. **Comets make regular paths around the sun, thus they can be used for telling seasons, days, and years.**

3. Many cultures, especially ancient cultures, viewed unusual activity in the heavens as bad omens, so the appearance of a comet would have been considered a bad omen. But God says not to be dismayed by them. Because they have regular orbits, their appearances can be predicted and there is nothing to fear.

4. There is no evidence for a comet nursery; comets were created in the beginning by God.

LESSON 10

METEORS

SHOOTING STARS

SUPPLY LIST

Pie pan   Salt   Flour   Marble   Toys   Golf ball

BEGINNERS

- What is a meteor? **A rock that is pulled into the earth's atmosphere and burns up.**
- Where do many meteors come from? **Broken up comets.**

WHAT DID WE LEARN?

- What is the difference between a meteoroid, meteor, and meteorite? **Meteoroids are small pieces of rock and other debris floating in space—usually orbiting the sun. Meteors are meteoroids that get close enough to the earth to be pulled in by the earth's gravity. Meteorites are meteors that reach the surface of the earth.**
- When is the best time to watch for meteors? **After midnight on any evening and especially around August 12 and November 17.**

TAKING IT FURTHER

- Space dust (extremely small meteorites) is constantly falling on the earth. If this has been going on for billions of years, what would you expect to find on the earth and in the oceans? **You would expect to find many meteorites in the fossil layers. You would also expect to find many feet of space dust accumulating in the oceans.**
- Have we discovered these things? **No. There have been very few confirmed meteorites found in the fossil layers, and a very small amount of space dust found in the oceans. Both of these facts indicate that the earth is relatively young.**

QUIZ 2

OUTER SPACE

LESSONS 5–10

Match the term with its definition.

1. _C_ Millions of stars rotating around a center
2. _L_ Name of our galaxy
3. _F_ Collection of planets orbiting the sun

4. _I_ A group of stars that form a picture

5. _B_ Star that doesn't move with respect to the earth's rotation

6. _A_ Unit of measurement for distances in space

7. _E_ An exploding star

8. _H_ Cloud of gas and dust in space

9. _K_ Scientific study of the universe/space

10. _M_ Superstitious belief that stars control the future

11. _D_ Gap between Mars and Jupiter

12. _G_ Balls of ice that orbit the sun

13. _J_ Piece of space debris that reaches the earth's surface

14. _O_ Piece of space debris that burns up in the atmosphere

15. _N_ What you can tell from a star's color

Short answer:

16. If a star has a blue color, is it hotter or cooler than our sun? **Hotter.**

17. What two things do we need to know to determine how far away a star is? **Brightness and amount of light emitted.**

18. Why do stars appear to move through the night sky? **Primarily earth's rotation (earth's revolution causes the starts to be in a different location from one night to the next).**

CHALLENGE QUESTIONS

Short answer:

19. Where do many evolutionists believe new stars are formed? Why is this unlikely? **In nebulae. Gas is expanding not contracting in nebulae.**

20. What names are given on a star map for the lines that are projected from the equator and the prime meridian? **Celestial equator and prime hour circle.**

21. What is the most common evolutionary explanation for the origins of the universe? **Big bang theory.**

22. Give one possible explanation for the ability to see distant starlight in a young world. **Earth is at the center of the universe and more time passed at the outer regions while the universe was expanding during creation than passed on earth. Another possible answer is that the speed of light was faster in the past.**

23. What is a group of asteroids traveling in the same path called? **A family.**

24. Name three asteroids in the Trojan Family. **Achilles, Hektor, Nestor, Agamemnon, Odysseus, Ajax, Diomedes, Antilochus, and Menelaus.**

25. Why does the existence of comets indicate that the universe is young? **Comets only exist for a few thousand years and we do not have any evidence that new comets are being formed.**

26. What is the most likely explanation for the extinction of dinosaurs? **Failure to adapt to changed climate after the Flood.**

UNIT 3
SUN & MOON

OVERVIEW OF OUR SOLAR SYSTEM

LESSON 11

REVOLVING AROUND THE SUN

SUPPLY LIST

A willingness to sing

Supplies for Challenge: Cardboard   String   2 thumb tacks   Paper

BEGINNERS

- How many planets are in our solar system? **8 (Pluto is no longer considered a planet).**
- What are the names of the small planets? **Mercury, Venus, Earth, Mars.**
- What are the names of the large planets? **Jupiter, Saturn, Uranus, Neptune.**

WHAT DID WE LEARN?

- Name the eight planets in our solar system. **Mercury, Venus, Earth, Mars, Jupiter, Saturn, Uranus, and Neptune—Pluto is no longer considered a planet, but a plutoid.**
- Name two of the dwarf planets. **Ceres, Pluto, and Eris were mentioned in this lesson.**
- Which planets can support life? **Only earth.**

TAKING IT FURTHER

- What are the major differences between the inner and outer planets? **Inner planets are closer to the sun, smaller, and are terrestrial (solid rock). Outer planets are larger and made from gas.**
- Why are the gas planets called Jovian planets? **Jovian means Jupiter-like. Jupiter is a gas giant. The other planets that are also comprised of gas, like Jupiter, are thus called Jovian.**

OUR SUN

LESSON 12

THE CENTER OF OUR SOLAR SYSTEM

SUPPLY LIST

Pie pan   Small mirror   Prism (optional)

Supplies for Challenge: Copy of "Sun Measurement" worksheet   2 index cards   Needle
Meter stick   Ruler   Tape   Calculator

BEGINNERS

- What is the sun? **A star in the center of our solar system.**

- What two things does the sun provide for the earth? **Heat and light.**

- How long does it take for the earth to go around the sun one time? **One year.**

WHAT DID WE LEARN?

- What are the main elements found in the sun? **95% of the sun is hydrogen and helium.**

- What colors are found in sunlight? **All colors from violet to red.**

TAKING IT FURTHER

- Why is the sun so important to us? **Its gravity holds everything in our solar system in place; it provides heat, light, and energy for life.**

- How does energy get from the sun to the earth? **It travels in waves—light, heat, radio, and x-rays.**

CHALLENGE: MEASUREMENT WORKSHEET

- What could account for the differences in your calculated value versus the known value? **The circle is small, so exact measurements are difficult to make. This causes error so your answer will differ from the expected value.**

- How does the diameter of the sun compare with the diameter of the earth? **The diameter of the earth is approximately 7,927 miles. The diameter of the sun is about 109 times bigger.**

LESSON 13 STRUCTURE OF THE SUN

WHAT IS IT LIKE ON THE INSIDE?

SUPPLY LIST

Sidewalk chalk   Note: It is most effective if you start this project in the morning.

BEGINNERS

- What are the two things the sun is made from? **Hydrogen and helium.**
- Why should you never look directly at the sun? **It can damage your eyes.**
- What are cooler areas of the sun called? **Sunspots.**

WHAT DID WE LEARN?

- What are the two parts of the sun's atmosphere? **The chromosphere and the corona.**

- What is a sunspot? **An area on the sun's surface that is cooler than surrounding areas.**

- Are sunspots stationary? **No, they tend to move from east to west across the surface of the sun.**

- What do scientists believe are the three parts of the sun's interior? **The core, radiative zone, and convective zone.**

TAKING IT FURTHER

- What is the hottest part of the sun? **The core, which is believed to be nearly 25 million degrees Fahrenheit.**

- What causes the aurora borealis or northern lights? **Particles emitted by a solar flare light up when they reach the earth's ionosphere.**

- When do you think scientists study the sun's corona? **The best time is during a solar eclipse. Scientists can see and study the corona while the rest of the sun is hidden by the moon.**

LESSON

14

SOLAR ECLIPSE

WHERE DID IT GO?

SUPPLY LIST

Flashlight   Tennis ball   Basketball or volleyball

Supplies for Challenge: Research materials on future lunar and solar eclipses

BEGINNERS

- What is a solar eclipse? **When the moon comes between the sun and the earth and blocks the sun's light from reaching the earth.**

- Does an eclipse block the light from the whole earth at once? **No, only a small area would be affected by an eclipse.**

- How often does an eclipse happen? **1 to 3 times a year.**

WHAT DID WE LEARN?

- What is an eclipse? **When one heavenly body blocks the light from another heavenly body.**

- What is the difference between a partial and a total eclipse? **A partial eclipse only covers part of the sun's disk, while a total eclipse covers it completely.**

- How often do solar eclipses occur? **1 to 3 times each year.**

TAKING IT FURTHER

- Why do you think plants and animals start preparing for nightfall during an eclipse? **Their instincts tell them that the sun is going down.**

- Why can a total eclipse only be seen in a small area on the earth? **The shadow of the moon is only about 150 miles across.**

- How can the moon block out the entire sun when the sun is 400 times bigger than the moon? **The moon is 400 times closer to the earth than the sun is, so from the earth they appear to be about the same size.**

LESSON 15

SOLAR ENERGY

CAN IT MEET OUR ENERGY NEEDS?

SUPPLY LIST

Copy of "Solar Energy" worksheet   Black, white, and green construction paper
2 clear glasses   Thermometer   Ice   Scissors and tape
Supplies for Challenge: Hardback books   Flashlight   Clipboard   Paper   Pencil

BEGINNERS

- What is solar energy? **Energy we get from the sun.**

- What are two ways that people use solar energy? **To heat water and to make electricity.**

SOLAR ENERGY WORKSHEET

- What color would you use to paint a solar collector? **Dark colors absorb more heat than light colors and dull surfaces absorb more heat than shiny ones. So solar collectors are painted with a dull black paint.**

WHAT DID WE LEARN?

- What is solar energy? **Energy we get from the sun.**

- What are the two ways that solar energy is used today? **To heat water and to generate electricity.**

TAKING IT FURTHER

- Why is solar energy a good alternative to fossil fuels? **It is clean, readily available, and virtually unlimited.**

- Why are the insides of solar collectors painted black? **Black absorbs the heat from the sun so black solar collectors are more efficient than other colors would be.**

- What are some of the advantages of using solar cells in outer space? **Other fuel sources would be too heavy to launch into space, but solar panels are very light. The sun shines all the time in space and there is no atmosphere to block the sun's energy.**

CHALLENGE: SOLAR ENERGY

- Is the new pattern bigger or smaller than the first pattern? **It should be bigger. The same amount of light is hitting the paper in both instances, but when the paper is at an angle, the light is more spread out, or dispersed.**

- Based on what you just learned, where would be the best location for a solar energy power plant? **The light hits the earth most directly at the equator, so a solar power plant would work best near the equator in an area that gets relatively few clouds.**

OUR MOON

IS IT MADE OF GREEN CHEESE?

SUPPLY LIST

Reflector (like one from a bicycle)   Flashlight

Supplies for Challenge: Binoculars or telescope (if available)

BEGINNERS

- Does the moon make its own light? **No.**
- Where does the moon's light come from? **The moon's light is light that is reflected from the sun.**
- How is the moon's surface different from the earth's surface? **It has no air and no plants or animals. It has many craters.**

WHAT DID WE LEARN?

- Why does the moon shine? **It reflects the light from the sun**.
- What causes the dark spots on the surface of the moon? **They are plains that are covered with hardened basalt.**

TAKING IT FURTHER

- Why does the size of our moon show God's provision for man? **It is much larger than most moons. This allows it to reflect a significant amount of light, lighting up the night.**
- Why is gravity much less on the moon than on the earth? **The moon is much less massive than the earth, and gravity is a function of mass.**
- Why doesn't the surface of the earth have as many craters as the surface of the moon? **Most meteors burn up in our atmosphere before they reach the earth's surface, but the moon does not have an atmosphere to protect it.**

MOTION & PHASES OF THE MOON

THERE'S A FULL MOON TONIGHT

SUPPLY LIST

Copy of "Identifying Phases of the Moon" worksheet

Supplies for Challenge: Copy of "Observing the Phases of the Moon" worksheet

BEGINNERS

- What two ways does the moon move? **Rotates on its axis, revolves around earth.**
- What is a new moon? **When the moon does not reflect any sunlight.**

IDENTIFYING PHASES OF THE MOON WORKSHEET

1. **New moon**   2. **Waxing crescent**   3. **First quarter**   4. **Waxing gibbous**
5. **Full moon**   6. **Waning gibbous**   7. **Last quarter**   8. **Waning crescent**

WHAT DID WE LEARN?

- What causes the phases of the moon? **The orbit of the moon around the earth causes it to be in a different position with respect to the sun each day.**

- Why does the same side of the moon always face the earth? **The moon rotates on its axis at the same rate that is revolves around the earth.**

- What causes a lunar eclipse? **The earth passes directly between the moon and the sun, blocking the light of the sun from the moon.**

- From the perspective of space, how long does it take for the moon to complete its cycle around the earth? **27.3 days, or about a month.**

TAKING IT FURTHER

- Why doesn't a lunar eclipse occur every month? **The moon's orbit is tilted 5 degrees from the earth's orbit so the sun, earth, and moon don't line up perfectly very often.**

- What is the difference between a waxing crescent and a waning crescent? **The waxing crescent is lit up on the right side of the moon from the earth's perspective, and is getting larger; the waning crescent is lit up on the left side of the moon, and is getting smaller.**

CHALLENGE: OBSERVING THE MOON

- When is the light side of the moon the same as the near side of the moon? **Full moon.**

- When is the dark side of the moon the same as the near side of the moon? **New moon.**

LESSON 18 ORIGIN OF THE MOON

WHERE DID IT COME FROM?

SUPPLY LIST

2 tops (spinning toys)   Masking tape
Supplies for Challenge: Bible

BEGINNERS

- Where did the earth come from? **God created it.**
- Where did the moon come from? **God created it, too.**
- On which day of creation did God create the moon? **Day 4.**

WHAT DID WE LEARN?

- What are four secular theories for the origin of the moon? **Capture Theory, Fission Theory, Accretion, and Impact Theory.**

- Which of these theories is most likely to be true? **None. They all have significant problems.**

- What does the Bible say about the origin of the moon? **It says that God created the moon on Day 4 to light up the night.**

TAKING IT FURTHER

- What are the main difficulties with the Capture Theory? **Scientists cannot explain what would cause the moon to leave its original orbit. The probability that the moon would approach the earth at exactly the right angle and speed to result in the moon orbiting the earth is extremely small.**

- Why do you think scientists come up with unworkable ideas for the moon's origin? **Scientists try to find answers to important questions. They come up with ideas for possible solutions. These are called hypotheses. They then test their hypotheses to see if they are true or not. Sometimes a hypothesis is wrong and should be cast aside. Generally, there are two problems with hypotheses concerning origins: 1. We cannot adequately test them because we cannot recreate the conditions under which the event took place; and 2. Many scientists are unwilling to consider the option that God exists, much less that He created things; therefore, many scientists still cling to unworkable ideas rather than admit that something outside of natural causes exists.**

CHALLENGE: ORIGIN OF THE MOON

- **Genesis 1:14–19 - God spoke the moon into existence on the fourth day of creation. Ps. 8:3–4 - The moon is the work of God's fingers; He set it in place. Ps. 33:6 - The heavens were made by God's word. Ps. 74:16 - God established the moon. Ps. 136:3–9 - God made the moon by His understanding (or wisdom). Jer. 31:35 - The Lord decreed the moon to shine.**

- **These verses clearly show that God created the moon out of nothing by speaking it into existence. They adequately explain the origin of the moon. The naturalistic explanations all have significant problems that cannot be explained by naturalistic means.**

QUIZ
3

SUN & MOON

LESSONS 11–18

List the planets in our solar system in order from the closest to the sun outward. **Mercury, Venus, Earth, Mars, Jupiter, Saturn, Uranus, Neptune.**

Mark each statement as either True or False.

1. _**F**_ A lunar eclipse occurs when the moon blocks the light from the sun.
2. _**T**_ The energy from the sun is generated by a process similar to a hydrogen bomb.
3. _**F**_ Scientists can directly view the interior of the sun.
4. _**F**_ The aurora borealis is a result of sunspots.
5. _**F**_ A total solar eclipse causes the whole earth to become dark.
6. _**T**_ Animals may act like night is falling during a solar eclipse.
7. _**T**_ It is very dangerous to look at the sun even during a total eclipse.
8. _**F**_ Solar collectors work best if they are painted glossy white.
9. _**T**_ Solar cells convert the sun's rays into electricity.
10. _**T**_ The same side of the moon always faces the earth.
11. _**F**_ Maria are areas on the moon filled with water.

12. _T_ The moon does not generate its own light.

Fill in the blank with the correct term.

13. The moon is called a **_full_** moon when it is on the opposite side of the earth from the sun.

14. The moon is called a **_new_** moon when it is on the same side of the earth as the sun.

15. The main elements found in the sun are **_hydrogen and helium_.**

CHALLENGE QUESTIONS

Match the term with its definition.

16. _B_ Squashed circle

17. _C_ Place in orbit closest to the sun

18. _G_ Place in orbit farthest from the sun

19. _J_ Center of a sunspot

20. _D_ Outer edge of a sunspot

21. _E_ Not concentrated

22. _F_ Plain filled with hardened basalt

23. _K_ Depression made by a meteorite

24. _I_ Valley on the moon

25. _A_ Side of the moon facing the earth

26. _L_ Side of the moon facing away from the earth

27. _H_ The side of the moon facing away from the sun

PLANETS

LESSON 19

MERCURY

CLOSEST PLANET TO THE SUN

SUPPLY LIST

Towel   Hair dryer   Ice
Supplies for Challenge: 2 index cards   Flashlight   2 clear plastic cups   Magnifying glass

BEGINNERS

- Name three things you learned about Mercury? **It is closest to the sun; it is the second smallest planet; it has very little atmosphere; it gets very hot and very cold.**

WHAT DID WE LEARN?

- How does Mercury's revolution around the sun and rotation on its axis compare to that of the earth? **Mercury travels quickly around the sun, but turns slowly on its axis compared to earth.**
- What is the surface of Mercury like? **Very hot or very cold, solid with lots of craters.**

TAKING IT FURTHER

- How does a lack of atmosphere affect the conditions on Mercury? **It causes extreme temperature swings and allows meteorites to strike the surface.**

LESSON 20

VENUS

THE SECOND PLANET

SUPPLY LIST

Copy of "Greenhouse Effect" worksheet   Shoebox   Plastic wrap   Aluminum foil   Tape
Thermometer
Supplies for Challenge: Shoebox   Modeling clay   Graph paper   Ruler   String   Washer

BEGINNERS

- What is the name of the second planet from the sun? **Venus.**
- Why couldn't you live on Venus? **It has a poisonous atmosphere and it is too hot.**

- Why can we see Venus in the night sky? **Its clouds reflect sunlight making it very bright.**

Greenhouse Effect worksheet

- What did you observe about the temperature in the box when it was covered with plastic wrap? **The temperature goes up inside the box.**
- Why did the temperature do this? **The plastic wrap traps some of the sun's rays.**
- What do you think the temperature would be inside the box if you left it in the sun for several hours? **The temperature will continue to rise for some time and then level off.**

What did we learn?

- Where is Venus's orbit with respect to the sun and the other planets? **It is second from the sun.**
- What makes Venus so bright in the sky? **Its atmosphere reflects the light of the sun.**
- What is a nickname for Venus? **The Morning Star or the Evening Star.**
- How many moons does Venus have? **None.**

Taking it further

- Even though Venus has an atmosphere, why can't life exist there? **The atmosphere is poisonous to people, plants, and animals. Also, it makes the planet too hot. The thick atmosphere also exerts too much pressure and would crush any living creatures.**
- Why doesn't the earth's atmosphere keep our planet too hot? **It is not nearly as thick as the atmosphere on Venus and it is composed of different gases. There is some concern about the greenhouse effect increasing on the earth because of increased carbon dioxide in the air. Some scientists feel that this is a real threat, while others feel that it is not. Most evidence points to higher carbon dioxide levels in the past, which may have actually been beneficial and not harmful. More study is needed in this area.**

LESSON 21

Earth

Designed for life

Supply list

1 orange per child   Markers   Globe of the earth   World map
Supplies for Challenge: 2 clear cups   Water   Milk   Flashlight

Beginners

- What is the third planet from the sun? **Earth.**
- List three ways that earth is just right for life. **Right distance from the sun, lots of water, good atmosphere.**
- How long does it take for the earth to make one trip around the sun? **1 year.**
- How long does it take for the earth to rotate once on its axis? **1 day.**

WHAT DID WE LEARN?

- What are some features of our planet that make it uniquely able to support life? **Just the right distance from the sun, axis is tilted just right to make most of the earth able to grow food, large amount of water, right atmosphere and weather patterns.**
- What name is given to the period of time it takes for the earth's revolution around the sun? **Year.**
- What name is given to the length of the earth's rotation on its axis? **Day.**
- On average, how far is the earth from the sun? **About 93 million miles or 150 million km.**

TAKING IT FURTHER

- What are some possible reasons why large amounts of water are found on earth but not on other planets? **Some planets are too hot and water evaporates away; some planets do not have enough gravity to hold an atmosphere so water also evaporates away; other planets do not have the right elements readily available; but most importantly, God created this planet uniquely for us.**
- Why is it important that earth is a terrestrial planet? **This means earth has a solid surface for us to live on.**

LESSON 22

MARS

THE RED PLANET

SUPPLY LIST

Empty aquarium or other case   Gloves   Liquid dish soap   Matches   Cup   Candle
Dry ice (Note: Dry ice must be obtained shortly before it is needed and must be handled by an adult with gloves.)

Supplies for Challenge: Research materials on Mars space probes

BEGINNERS

- What is the name of the fourth planet from the sun? **Mars.**
- How big is Mars compared to earth? **About half as big.**
- Why is Mars sometimes called the red planet? **It has rust in its soil giving it a red tint.**

EXPERIMENTING WITH POLAR ICE CAPS

- What was the "smoke" coming off of the dry ice? **Carbon dioxide gas.**
- Why did the candle flame go out? **The carbon dioxide in the cup was heavier than air and pushed the oxygen away from the candle so the flame went out.**
- Why did the water in the cup "boil"? **As the dry ice melted it quickly turned to gas, which bubbled to the surface of the water.**

WHAT DID WE LEARN?

- Why is Mars called a superior planet? **It has an orbit that is larger than the earth's orbit.**
- Why is Mars called the red planet? **Its soil has a red color due to a high amount of iron oxide—rust.**
- How many moons does Mars have? **2—Phobos and Deimos.**

- What causes the dust storms on Mars? **Heat from the sun causes winds that pick up dust**.

- Why doesn't the wind on earth cause giant dust storms like the wind on Mars? **Although heat from the sun does cause wind, most of the earth is covered with water. Also, the land is mostly covered with vegetation that holds down the soil, so giant dust storms are unlikely. Small dust and sand storms do happen on earth.**

- How would your weight on Mars compare to your weight on Mercury? **They would be about the same because the gravity it about the same on both planets.**

LESSON 23

JUPITER

THE GAS GIANT

SUPPLY LIST

2 cereal bowls   Marbles—enough to fill both bowls

Supplies for Challenge: Clear cup   Water   Tea bag   Pencil

BEGINNERS

- What is the name of the fifth planet? **Jupiter.**

- Name three things that are special about Jupiter. **It is made of gas not rock, it is the largest planet, it has a Great Red Spot, it has more than 60 moons.**

WHAT DID WE LEARN?

- What are some major differences between Jupiter and the inner planets? **The inner planets are solid and relatively small. Jupiter is very large and made of gas.**

- What is the Great Red Spot? **It is believed to be a giant wind storm that has lasted hundreds of years.**

TAKING IT FURTHER

- Why does Jupiter bulge more in the middle than earth does? **Jupiter spins faster on its axis than the earth does, causing more outward or centrifugal force.**

- Why can't life exist on Jupiter? **The temperatures are too cold, the surface is not solid, there is no air to breathe.**

- Why are space probes necessary for exploring other planets? **Probes can go where people can't. They can see things up close that we cannot see from earth. For example, Voyager discovered a ring around Jupiter. Also, probes can go into environments that are difficult for humans to enter such as Venus's carbon dioxide and sulfuric acid atmosphere. Probes can do tests in far away places like the soil tests done on Mars by the Spirit and Opportunity rovers.**

LESSON 24

SATURN

SURROUNDED BY BEAUTIFUL RINGS

SUPPLY LIST (none)

BEGINNERS

- What are Saturn's rings made from? **Ice, dust, and rocks.**
- What is the name of Saturn's largest moon? **Titan.**

WHAT DID WE LEARN?

- Who first saw Saturn's rings? **Galileo; however he did not recognize them as rings. Christian Huygens first identified the rings.**
- What are Saturn's rings made of? **Pieces of ice, dust, and rock.**
- What makes Titan unique among moons? **It is the only moon with an atmosphere.**

TAKING IT FURTHER

- Why did astronomers believe that Saturn had only a few rings before the Voyager space probe explored Saturn? **That was all that could be observed from earth using telescopes.**
- Both Titan and earth have a mostly nitrogen atmosphere. What important differences exist between these two worlds that make earth able to support life but Titan unable to? **Titan is too far away from the sun, and thus is too cold to support life. Also, Titan's atmosphere has methane and no oxygen. Earth has just the right amount of oxygen to support life.**

LESSON 25

URANUS

SEVENTH PLANET FROM THE SUN

SUPPLY LIST

Ping-pong ball   Paint   Basketball, volleyball, or other larger ball
Supplies for Challenge: Modeling Clay   8 pencils   Protractor   Index cards

BEGINNERS

- What is the name of the seventh planet from the sun? **Uranus.**
- What color is Uranus? **Pale blue.**
- How does Uranus move that is different from other planets? **It rolls around the sun instead of spinning upright.**

WHAT DID WE LEARN?

- What makes Uranus unusual compared to the other planets? **It rotates on its side.**
- How have rings been discovered around Uranus? **By telescopes and space probes.**

TAKING IT FURTHER

- How can we learn more about Uranus? **Send out more probes; build better telescopes.**
- Why is Uranus such a cold planet? **It is too far from the sun for the sun to heat it very much.**

NEPTUNE

LAST OF THE GAS GIANTS

SUPPLY LIST

Flashlight   3 clear plastic or glass cups   Red and blue food color
Supplies for Challenge: String   Washer

BEGINNER

- What is the name of the eighth planet from the sun? **Neptune.**
- What color is Neptune? **Blue.**
- Is Neptune a solid or a gas planet? **Gas.**

WHAT DID WE LEARN?

- What similarities are there between Uranus and Neptune? **They are both gas planets and have rings and moons. They both have methane in their atmospheres that make them appear blue. They were both discovered in the last 250 years.**
- What are two possible explanations for the Great Dark Spot? **Some scientists believe it was a giant windstorm in Neptune's atmosphere, others believe it may have been a hole in the clouds surrounding the planet.**

TAKING IT FURTHER

- Explain how Neptune was discovered. **Observations of Uranus's orbit indicated that there must be a planet whose gravity was affecting Uranus. Astronomers/mathematicians calculated where the unknown planet had to be and another astronomer found it there.**
- What affects the color of a planet? **Many things can affect what color a planet appears to be. Mars is red because of the rust (iron oxide) in its soil; earth appears blue because of the water on the surface and its atmosphere; Uranus and Neptune appear blue because of the methane in their atmospheres.**

Pluto & Eris

Dwarf planets

Supply list

Copy of "How Much Do I Weigh?" worksheet   Calculator   Bathroom scale

Beginner

- Is Pluto considered a true planet? **No, it is now classified as a dwarf planet.**
- What is the temperature like on Pluto? **Very cold.**
- How many known moons does Pluto have? **Three.**
- What is the name of the largest moon orbiting Pluto? **Charon.**

What did we learn?

- What was the last planet to be discovered in our solar system? **Pluto; there may be one or more planets farther out than Pluto.**
- How does the gravity on Pluto compare to the gravity on earth? **There is practically no gravity on Pluto. It is 0.08 times the gravity on earth.**
- Is Pluto always farther away from the sun than Neptune? **No, it is closer to the sun than Neptune for 20 out of every 250 years.**
- What is unique about how Charon orbits Pluto? **They are in a synchronous orbit—the same sides of the planet and moon always face each other**.

Taking it further

- Why did it take so long to discover Pluto? **It's a small planet and very far away. It was much dimmer than expected**.
- Why is Pluto no longer considered to be a planet? **It is too small; it does not meet the definition for a planet adopted by the International Astronomical Union.**
- What alternate classification was given to Pluto in 2006? **It was called a dwarf planet.**

Planet	Terrestrial or Jovian	Atmosphere (Yes/No) If yes, what is it made of?	# of known moons	Rings (Yes/No)	Surface temp (Hot/Cold/Comfortable)
Mercury	Terrestrial	Not much—very thin helium/hydrogen	0	No	Hot and cold
Venus	Terrestrial	Yes—carbon dioxide/ nitrogen/sulfuric acid	0	No	Hot
Earth	Terrestrial	Yes—nitrogen/oxygen	1	No	Comfortable
Mars	Terrestrial	Yes—carbon dioxide	2	No	Comfortable to cold
Jupiter	Jovian	Yes—hydrogen	60+	Yes	Cold
Saturn	Jovian	Yes—hydrogen/helium	30+	Yes	Cold
Uranus	Jovian	Yes—hydrogen/helium/ methane	20+	Yes	Cold
Neptune	Jovian	Yes—hydrogen/helium/ methane	13	Yes	Cold
Pluto (dwarf planet)	Neither/ unknown	Possibly/Sometimes— very thin methane	3	No	Cold

CHALLENGE QUESTIONS

- **Mercury—has a slight atmosphere, has a magnetic field, is similar to our moon; Venus—surface geography, temperature and pressure; Earth—not needed; Mars—Soil composition, surface geography, possible water ice; Jupiter—information on Great Red Spot; Saturn—shepherd moons, Enceladus's role in ring formation; Uranus and Neptune—discovery of moons; Pluto—none yet.**

SPACE PROGRAM

NASA

THE NATIONAL AERONAUTICS AND SPACE ADMINISTRATION

SUPPLY LIST

Tagboard/poster board   Steel BBs   Magnet   Plastic lid or dish   Several books

BEGINNERS

- What is NASA? **The National Aeronautics and Space Administration—a group of people who work on different ways to study space.**

- Name three ways that NASA studies things in space. **Launching rockets, Space Shuttle, Space Station, Hubble Telescope, by conducting experiments, and training astronauts.**

- What was the first important job that NASA had to do? **Find a way to send a man to the moon.**

WHAT DID WE LEARN

- What is NASA? **The National Aeronautics and Space Administration, a science organization for studying the universe.**

- When was NASA formed? **1958.**

- What was one of NASA's first tasks? **To put a man on the moon.**

- List at least three of NASA's work groups. **Improving aeronautics, space flight, space probes, robotics, space station, analyzing data, planning missions.**

TAKING IT FURTHER

- How does NASA help people who are not interested in space exploration? **By developing technology that is applicable in other areas.**

- How might an evolutionary worldview affect NASA's work? **Many of NASA's projects are dedicated to finding life on Mars or other planets/moons; others are designed to prove the Big Bang. If NASA had a biblical worldview, their missions could be discovering the wonders of the universe that glorify God.**

CHALLENGE: NACA

- What was NACA? **National Advisory Committee for Aeronautics.**

- What was its original purpose? **To supervise and direct the scientific study of the problems of flight, with a view to their practical solution.**

- What were some of the major contributions to aeronautics that were made by NACA? **Supersonic and hypersonic flight technology, safety designs, improved engines, airfoils, and wings.**

SPACE EXPLORATION

SEEING WHAT'S OUT THERE

SUPPLY LIST

Styrofoam balls   Toothpicks   Empty thread spools   Aluminum foil   Modeling clay
Tagboard/poster board   Model rocket and launch pad (optional)
Supplies for Challenge: Drawing materials

BEGINNERS

- What is a satellite? **Something that orbits the earth.**

- What is a space probe? **Something sent to other planets to take pictures and get other information.**

- What invention was needed in order to send satellites and other items into space? **Rockets.**

WHAT DID WE LEARN?

- Who were the first people to talk about going into space? **The science fiction writers of the 19th century.**

- Who is considered the father of modern rocketry? **Robert H. Goddard.**

- What major event sparked interest in the development of the rocket for space travel? **The extensive use of rockets during World War II.**

- Who was one of the primary developers of rockets in the United States after World War II? **The German scientist, Werner VonBraun.**

- What was the first man-made object to orbit the earth? **Sputnik—a satellite launched by the Soviet Union.**

- Who was the first man in space? **Yuri Gagarin.**

- Who was the first American in space? **Alan Shepherd.**

- Who was the first American to orbit the earth? **John Glenn.**

- Who was the first man to walk on the moon? **Neil Armstrong.**

TAKING IT FURTHER

- Why are satellites an important part of space exploration? **Satellites have many purposes including collecting scientific and military data, communications, and navigation.**

- Why are space probes an important part of space exploration? **Probes can go to places that are too far away for humans to travel and places that are too dangerous for humans. The atmosphere on Venus crushed several of the early probes that were sent there. Losing a probe is a risk worth taking, but risking human life is not. Most of the information we have about the other planets has come from space probes.**

LESSON 30

APOLLO PROGRAM

FIRST FLIGHT TO THE MOON

SUPPLY LIST

String (enough to reach across a room)   2 balloons   2 straws   Tape

BEGINNERS

- What was the name of the project that sent men to the moon? **Apollo.**
- How many stages or rocket engines did the Saturn V rocket have? **Three.**
- Who were the first men on the moon? **Neil Armstrong and Buzz Aldrin.**

WHAT DID WE LEARN?

- What was the name of the NASA program whose goal was to put a man on the moon? **Apollo.**
- What are the three modules in the Apollo spacecraft? **The command module, the service module, and the lunar module.**
- What were the two parts of the lunar module designed to do? **The descent stage allowed astronauts to land on the moon. The ascent stage lifted them from the moon back to the command module in lunar orbit.**
- What was the name of the three-stage rocket used with the Apollo spacecraft? **Saturn V.**

TAKING IT FURTHER

- What is the advantage of a multi-stage rocket engine? **The first engine must lift all of the weight of the combined system, but the second engine only needs to lift the weight of the system after the first stage is gone, so it does not need to be as big. The third stage is only needed to break the modules out of earth's orbit, so it can be relatively small.**

LESSON 31

THE SPACE SHUTTLE

REUSABLE PARTS

SUPPLY LIST

Copy of "Space Shuttle" worksheet

BEGINNERS

- What is the space shuttle? **A reusable space ship for orbiting the earth.**
- What are the three parts of the space shuttle system? **The orbiter, the fuel tank, and the solid rocket boosters.**
- What piece of equipment is in the payload bay that is helpful to astronauts? **A robotic arm.**

SPACE SHUTTLE WORKSHEET

See drawing in student manual, page 137.

WHAT DID WE LEARN?

- What is the main advantage of the space shuttle vehicle over all previous manned space vehicles? **The shuttle is reusable, thus it is much less expensive to operate.**

- What are the main purposes of the shuttle program? **Scientific experiments, launching of space satellites, and ferrying astronauts and supplies to the space station.**

- What are the two main parts of the orbiter and what are their purposes? **The crew cabin contains the flight deck and living areas. The payload bay provides room for satellites and experiments. It also provides an area where repair work can be done**.

TAKING IT FURTHER

- Why is the space shuttle called an orbiter? **It is designed to orbit the earth for experimental purposes. It is not designed for outer space flight.**

- Why is the orbiter shaped like an airplane? **The shape of a space vehicle is relatively unimportant in space because there is no gravity and no atmosphere. However, the shuttle must be able to land safely on earth. So it is designed with aerodynamics similar to an airplane, so it can land like a plane in the earth's atmosphere.**

- Why does the orbiter have to be carried back to Florida if it lands in California? **The shuttle does not have any jet engines or any way to propel itself through the atmosphere. It only has booster engines that allow it to move in space. So, although it may resemble an airplane, it can't fly like one.**

LESSON 32
INTERNATIONAL SPACE STATION

REACHING FOR FREEDOM

SUPPLY LIST

Water   Waxed paper   Toothpick

BEGINNERS

- What is the name of the current space station? **International Space Station.**

- How does the space station get power? **From large solar panels that convert the sun's energy into electricity.**

- Why can astronauts do different experiments on the space station than on earth? **There is very little gravity on the space station.**

WHAT DID WE LEARN?

- What is the International Space Station? **A permanent orbiting laboratory in space.**

- Why do countries feel there is a need for a space station? **To study long-term effects of micro-gravity for scientific purposes and to develop new technologies to benefit all of humanity.**

- What shape would you expect a flame to be on the space station? **A candle flame is somewhat teardrop shaped on earth. However, in space it is circular because the oxygen molecules are equally available from all directions and are not being pulled down by gravity.**

LESSON 33 ASTRONAUTS

MODERN DAY EXPLORERS

SUPPLY LIST

Winter clothing including hat, gloves, coat, snow pants, and boots   Hand mirror

Building blocks   Nut and bolt   Optional: Bicycle helmet or motorcycle helmet with face mask

Supplies for Challenge: Research materials on astronauts

BEGINNERS

- What are two important subjects to study in school if you want to become an astronaut? **Math and science.**
- What things do space suits provide that are missing in space? **Air, air pressure, water, heating and cooling, the ability to communicate.**
- How does an astronaut move around in space? **With a rocket pack.**

WHAT DID WE LEARN?

- What are some ways that astronauts train for their missions? **They learn about the vehicles they will be using and the experiments they will be performing. They practice in their spacesuits underwater. They ride in the "Vomit Comet."**
- What conditions in space require astronauts to need spacesuits? **There is no atmosphere in space so there is no pressure, no oxygen, and no protection from the hot and cold extremes in space. Also, there is more radiation in space so extra protection is needed.**

TAKING IT FURTHER

- What are some things you can do if you want to become an astronaut? **Study math and science, keep physically fit, and work hard.**
- What would you like to do if you were involved in the space program? **Answers will vary.**

QUIZ 5 SPACE PROGRAM

LESSONS 28–33

Choose the best answer for each question.

1. _B_ At this time, what is the best way to study long-term effects of zero-gravity?

2. _A_ Which characteristic is generally not a quality of an astronaut?

3. _D_ What was designed to protect astronauts in space?

4. _C_ What wartime invention led to space exploration?

5. _A_ What object was launched into space on Oct. 4, 1957?

6. _B_ Who was the first man in space?

7. _C_ Who challenged America to put a man on the moon before 1970?

8. _A_ Which of the following is not a function of space satellites?

9. _D_ Which of the following programs did not help to put a man on the moon?

10. _B_ What rocket was used in the Apollo space program?

11. _C_ Who was the first person to walk on the moon?

12. _C_ Which of the following was not left on the moon?

13. _D_ What shape is the shuttle orbiter?

14. _B_ What is the maximum number of crew members on the space shuttle?

CHALLENGE QUESTIONS

Short answer:

15. List three ways that NACA helped improve flight. **Improved air foils, engines, and wings, safety improvements, ice reduction processes, hypersonic and supersonic designs.**

16. List three challenges unique to living and working in space. **No gravity, no air, no air pressure, no heat and cold protection, radiation.**

17. Why is private space research important? **Competition spurs innovation; commercial applications will help support further research; improvements in space research can be applied to other areas of life; private space research looks at areas that are not funded by the government so improvements are made in more areas.**

18. List two differences between the space shuttle and the Orion. **Shuttle is shaped like a plane while Orion is shaped like a capsule; shuttle is only an orbiter while Orion will be able to go to the moon; shuttle can land on an airstrip while Orion must land in the water.**

19. List two ways that the space shuttle and the Orion are similar. **Both are reusable; both can take people to and from the space station; both can launch satellites.**

20. What is one fascinating thing you have learned? **Answers will vary.**

LESSON 34
SOLAR SYSTEM MODEL: FINAL PROJECT

SHOWING WHAT'S OUT THERE

FINAL PROJECT SUPPLY LIST

Styrofoam balls in the following sizes: 5 in. (1 each)   4 in. (1 each)   3 in. (1 each)
2½ in. (1 each)   2 in. (1 each)   1½ in. (2 each)   1¼ in. (2 each)
2 Styrofoam 4½-inch rings   5 stiff craft wires—each 14 inches long   Paint

Supplies for Challenge: Index cards

WHAT DID WE LEARN?

- What holds all of the planets in orbit around the sun? **The force of gravity.**
- What other items are in our solar system that are not included in your model? **Asteroids, comets, meteoroids, Pluto and other dwarf planets.**

TAKING IT FURTHER

- Why do the planets orbit the sun and not the earth? **The sun is the most massive object in the solar system. It therefore has the strongest gravitational pull so smaller items, such as planets, will orbit it.**

OUR UNIVERSE

LESSONS 1–34

Fill in the blank with the correct term from below.

1. An _asteroid_ is a chunk of rock in a regular orbit around the sun.
2. The surface of the moon is covered with dark areas called _maria_.
3. A piece of space debris that burns up in the earth's atmosphere is a _meteor_.
4. Space _probes_ can explore areas that man cannot.
5. A piece of space debris that hits the earth's surface is a _meteorite_.
6. A _supernova_ is a star that experiences a very large explosion.
7. A _comet_ is a ball of ice and dust that orbits the sun.
8. A star that has exploded and collapsed in on itself is called a _black hole_.
9. An _eclipse_ occurs when one heavenly body blocks the light from another heavenly body.
10. The gases surrounding a planet are its _atmosphere_.
11. A large cloud of gas and dust in space is a _nebula_.
12. Heated plasma that extends from the surface of the sun to 6,200 miles is the _chromosphere_.
13. The _corona_ is the outermost part of the sun's atmosphere.
14. The visible surface of the sun is called the _photosphere_.
15. _Solar energy_ is energy from the sun.
16. A _satellite_ is anything that has a regular orbit around a planet.

Short answer:

17. List at least one unique characteristic for each planet (or plutoid). **Accept any reasonable answer.**
    Mercury—Closest to the sun, little or no atmosphere, no moons, extreme temperatures.
    Venus—Sulfuric acid clouds, hottest planet in the solar system, closest in size to earth.
    Earth—Only planet with life, significant amount of water, designed by God for us.
    Mars—Red planet, frozen carbon dioxide at the poles, most space probes.
    Jupiter—Largest planet, Great Red Spot, gas giant.
    Saturn—Beautiful rings, second largest planet, gas giant.
    Uranus—Rotates on its side, blue color.
    Neptune—Methane atmosphere gives it a blue color, had the Great Dark Spot.

Pluto—Former planet, classified as a dwarf planet, Charon in synchronous orbit.

18. Describe why gravity is important to our solar system. **Gravity holds everything in place. It makes the planets orbit the sun. It makes the moons orbit the planets. It holds our atmosphere in place.**

19. Place these colors of stars in order from coolest to hottest: blue, orange, yellow, white: **Orange, yellow, white, blue.**

20. What are the two ways that planets move through space? **Rotate on axes, revolve around the sun.**

21. List three tools used to study space. **Space probe, space ships/shuttle, space station, satellite, telescopes.**

22. What was the purpose of the Apollo missions? **To send a man to the moon.**

23. Why was the space shuttle developed? **To make a reusable space ship, to do research in orbit.**

24. What is the purpose of the International Space Station? **To have a place to conduct long-term micro-gravity experiments.**

25. List three purposes of a space suit. **To protect the astronaut from harmful radiation and extreme temperature, to provide pressure and air, to provide communications.**

CHALLENGE QUESTIONS

Mark each statement as either True or False.

26. _T_ A Foucault pendulum demonstrates the rotation of the earth.

27. _F_ Kepler was the first to suggest the heliocentric model of the universe.

28. _F_ Distant starlight proves the universe is billions of years old.

29. _T_ The celestial equator on a star map corresponds to the equator on a map of the earth.

30. _T_ Sir Isaac Newton improved on Galileo's design of the early telescope.

31. _T_ The larger the opening of a telescope, the more you can magnify the image.

32. _T_ Hektor and Achilles are two asteroids in the Trojan asteroid family.

33. _F_ Science has proven that a meteor led to the extinction of the dinosaurs.

34. _T_ Kepler's laws of planetary motion explain why planets move in ellipses.

35. _F_ Solar flares are unrelated to sunspots.

36. _F_ All planets are at the same tilt with respect to the sun.

37. _T_ There are four Jovian and four terrestrial planets.

38. _T_ Mercury and Venus have very different atmospheres from earth.

39. _T_ More space probes have gone to Mars than to any other planet.

40. _F_ SpaceShipOne proved that private space research is unrealistic.

41. _T_ Orion is being modeled after the Apollo program.

42. _F_ Only official astronauts are allowed on the International Space Station.

43. _T_ Most astronauts have had a military background.

CONCLUSION

REFLECTING ON OUR UNIVERSE

SUPPLY LIST

Bible   Flashlight   Blanket   Night sky

WHAT DID WE LEARN?

• What is the best thing you learned about our universe? **Answers will vary.**

RESOURCE GUIDE

Many of the following titles are available from Answers in Genesis (www.AnswersBookstore.com).

Suggested Books

Glow-In-The-Dark Nighttime Sky by Clint Hatchett—Easy-to-use star charts

Astronomy for Every Kid by Janice VanCleave—Many fun activities

The Astronomy Book by Jonathan Henry—a wealth of knowledge on subjects such as supernovas, red shift, facts about planets and much more

Astronomy and the Bible: Questions and Answers by Donald DeYoung—Answers to 110 questions on astronomy and the universe

Our Created Moon by Don DeYoung & John Whitcomb—Answers to 63 questions about the "lesser light"

Universe by Design by Danny Faulkner—Explores and explains the historical development of the science of astronomy from a creationist view

Taking Back Astronomy by Dr. Jason Lisle— Christian apologetics for astronomy

Suggested Videos

Newton's Workshop by Moody Institute— fun live action videos with Christian themes

Journey to the Edge of Creation by Moody Institute— beautiful film of universe

Creation Astronomy: Viewing the Universe Through Biblical Glasses by Dr. Jason Lisle (DVD)— Shows how the evidence of nature lines up perfectly with the clear teachings of Scripture

Created Cosmos: A Creation Museum Planetarium Show by Dr. Jason Lisle (DVD)—A visually stimulating tour of the universe underscoring its incomprehensible size and structure

Field Trip Ideas

- Creation Museum in Petersburg, Kentucky
- Observatory
- Space center or Space museum
- Planetarium
- Drive out to the country, away from city lights, to observe the night sky

CREATION SCIENCE RESOURCES

Answers Book for Kids Four volumes by Ken Ham with Cindy Malott—Answers children's frequently asked questions

The New Answers Books 1 & 2 by Ken Ham and others—Answers frequently asked questions

The Amazing Story of Creation by Duane T. Gish—Gives scientific evidence for the creation story

Creation Science by Felice Gerwitz and Jill Whitlock—Unit study focusing on creation

Creation: Facts of Life by Gary Parker—Comparison of the evidence for creation and evolution

The Young Earth by John D. Morris—Lots of facts disproving old-earth ideas

MASTER SUPPLY LIST

The following table lists all the supplies used for *God's Design for Heaven & Earth: Our Universe* activities. You will need to look up the individual lessons in the student book to obtain the specific details for the individual activities (such as quantity, color, etc.). The letter *c* denotes that the lesson number refers to the challenge activity. Common supplies such as colored pencils, construction paper, markers, scissors, tape, etc., are not listed.

Supplies needed (see lessons for details)	Lesson
Aluminum foil	20, 29
Aquarium or other empty case	22
Balloons	30
Basketball or volleyball	3, 14, 25
Bathroom scale	27
Bible	1, 18c, 35
Building blocks	33
Calculator	6c, 27
Candle	22
Cardboard	11c
Cereal bowls	23
Clipboard	15c
Craft wire	34
Cups (clear plastic or glass)	15, 19c, 21c, 22, 23c, 26
Dry ice	22
Flashlight	3, 4, 6, 7, 14, 15c, 16, 19c, 21c 26, 35
Flour	10
Food coloring	26
Glitter	9
Globe of the earth	21
Gloves	22
Golf ball	2, 10
Graph paper	20c
Hairdryer	19
Ice	15, 19
Index card	12c, 19c, 25c, 34c
Liquid dish soap	22

Supplies needed (see lessons for details)	Lesson
Magnet	28
Magnifying glass	4, 19c
Marbles	10, 23
Masking tape	3, 18
Matches	22
Milk	21c
Mirror	4, 12, 33
Model rocket and launch pad (optional)	29
Modeling clay	3c, 20c, 25c, 29
Motorcycle helmet with face plate, or bike helmet (optional)	33
Nut and bolt	33
Orange (fruit)	21
Paint	25, 34
Pencils (wooden)	25c
Ping-pong ball	2, 25
Plastic lid or dish	28
Plastic wrap	20
Plastic zipper bag	
Poster board/tagboard	9, 28, 29
Prism (optional)	12
Protractor	25c
Reflector (like from a bicycle)	16
Ruler	6, 20c
Salt	10
Shoe box	20, 20c
Sidewalk chalk	13
Star chart	5
Steel BBs	28
Straw	30

Supplies needed (see lessons for details)	Lesson
String	11c, 20c, 26c, 30
Styrofoam balls	9, 29, 34
Styrofoam rings	34
Tea bag	23c
Telescope (optional)	4, 16c
Tennis ball	14
Thermometer	15, 20
Thumb tacks	11c
Tops (spinning toys)	18

Supplies needed (see lessons for details)	Lesson
Towel	19
Toy houses, cars, etc.	10
Tripod	3c
Turntable (swivel chair, stool, Lazy Susan, etc.)	3c
Washer	20c, 26c
Waxed paper	32
Winter clothes	33
World atlas/map	21
Yard stick/meter stick	6, 12c

WORKS CITED

"About Foucault Pendulums and How They Prove the Earth Rotates!" http://www.calacademy.org/products/pendulum.

"About NASA." http://www.nasa.gov/about/highlights/index.html.

Arty Facts Space & Art Activities. Ed. Ellen Rodger. New York: Crabtree Publishing Company, 2002.

Behrens, June. *Sally Ride, Astronaut An American First*. Chicago: Children's Press, 1984.

Bonnet, Bob, and Dan Keen. Flight, Space & Astronomy. New York: Sterling Publishing, 1997.

Bourgeois, Paulette. *The Sun*. Buffalo: Kids Can Press, Ltd., 1997.

Caprara, Giovanni. *Living in Space*. Milan: Firefly Books, 2000.

"Civilian Space Travel." http://www.kidsastronomy.com/civilian_space_travel.

Cole, Michael D. *Hubble Space Telescope Exploring the Universe*. Springfield: Enslow Publishers, Inc., 1999.

Cole, Michael D. *NASA Space Vehicles*. Berkley Heights: Enslow Publishers, Inc., 2000.

"Deep Impact." http:// deepimpact.umd.edu.

DeYoung, Donald. *Astronomy and the Bible*. Grand Rapids: Baker Book House, 1989.

Dickinson, Terrance. *Exploring the Night Sky*. Toronto: Camden House, 1987.

"Finding the Size of the Sun and the Moon." http://cse.ssl.berkeley.edu/AtHomeAstronomy/activity_03.html

"First Flight of SpaceShipOne Into Space." http://www.richard-seaman.com/Aircraft/AirShows/SpaceShipOne2004/.

Gifford, Clive. *The Kingfisher Facts and Records Book of Space*. New York: Kingfisher, 2001.

Hatchett, Clint. *The Glow-in-the-Dark Night Sky Book*. New York: Random House, 1988.

Henry, Jonathan. *The Astronomy Book*. Green Forest: Master Books, 2005.

Hitt, Robert, Jr. *The Sun's Family*. Danbury: Grolier Educational, 1998. Vol. 1 of *Outer Space*.

"James Webb Space Telescope." http://ngst.gsfc.nasa.gov.

Kerrod, Robin. *The Moon*. Minneapolis: Lerner Publications Co., 2000.

"Jupiter's New Red Spot." http://science.nasa.gov/headlines/y2006/02mar_red.

"Mauna Kea Telescopes." http://www.ifa.hawaii.edu/mko/telescope_table.htm.

Meyers, Robert. "Giant Telescopes Combine to Form World's Largest." http://www.space.com/scienceastronomy/astronomy.

National Aeronautics and Space Administration. *The Amazing Hubble Space Telescope*. John F. Kennedy Space Center: NASA, 1986.

"New Horizons." http://www.nasa.gov/mission_pages/newhorizons/main/.

"Prepared for the Mission: A Tribute to Rick Husband." *Homeschooling Today*. Mar/Apr 2003: 30–33.

Rau, Dana M. *Jupiter*. Minneapolis: Compass Point Books, 2002.

Rau, Dana M. *Mars*. Minneapolis: Compass Point Books, 2002.

Rau, Dana M. *Mercury*. Minneapolis: Compass Point Books, 2002.

Rau, Dana M. *Neptune*. Minneapolis: Compass Point Books, 2003.

Rau, Dana M. *Pluto*. Minneapolis: Compass Point Books, 2003.

Rau, Dana M. *Saturn*. Minneapolis: Compass Point Books, 2003.

Rau, Dana M. *Uranus*. Minneapolis: Compass Point Books, 2003.

Rau, Dana M. *Venus*. Minneapolis: Compass Point Books, 2002.

"Raymond Orteig—$25,000 Prize." http://www.charles-lindbergh.com/plane/orteig.asp.

"Saturn Probe Sights Mystery Moon." http://news.bbc.co.uk/1/hi/sci/tech/3633297.stm.

Simon, Seymour. *Our Solar System*. New York: Morrow Junior Books, 1992.

"Saturn's Rings." http://saturn.jpl.nasa.gov/science/index.cfm?PageI.

"Space Camp." http:// www.spacecamp.com.

Spangenburg, Ray, and Kit Moser. *Mercury*. New York: Franklin Watts, 2001.

Spangenburg, Ray, and Kit Moser. *Venus*. New York: Franklin Watts, 2001.

"Sunspots." http://www.exploratorium.edu/sunspots/research7.html.

VanCleave, Janice. *Astronomy for Every Kid*. New York: John Wiley & Sons, Inc., 1991.

VanCleave, Janice. *Solar System*. New York: John Wiley& Sons, 2000.

"The Vision for Space Exploration." http://www.nasa.gov/missions/solarsystem/explore_main_old.html.